WROUGHT IRONWORK, WELDING & STEEL FABRICATION VOL 1

Alex Mayo

COPYRIGHT

Acknowledgements:

Zac Bull – _Class 1 welder_, Terry Cody - _Gate Maker_

Katie Mae– _Admin_   Sian Toni- _Site Survey and Design_

Introduction

Hopefully this Book has grabbed your attention and you may feel that you have or can acquire and learn the necessary practical skills to make the dream of working for yourself a reality, or maybe you're an existing tradesperson wanting to aquire extra skills? Or at the least you are considering making some projects for your home and garden? In either case you will find this Training Manual Invaluable. I intend to publish a follow up book(VOL2), based on general fabrication with examples of work and also other technical skills such as plate development, Tig Welding, Sheetmetalwork, Pressbrake work, Guillotine, Punch Press etc...

Written originally and partly for company trainees, my aim is to bring this knowledge into the realms of both the Amateur Hobbyist,Professional Craftsman and small businessman wishing to take advantage of the ever growing market of Ornamental iron Gates, Railings etc.....

With two accomplished time served tradesmen doing the writing, with 50 years experience between them who have worked in all aspects of the Fabrication business from shopfloor to Director level, and experienced with dealing and selling to Homeowners, Builders, Property Developers, Gardening Centres as well as through various selling medias such as Ebay or via a website.

You will find the training and tips laid out in laymans terms through experience in the field and to some extent 'trial and error'.

I've included a description of equipment that would be required and a detailed description of useful jigs, welding methods and measuring up tips.

Information and examples of How to cost,Invoice, purchase etc.... Details on how to estimate and quote successfully with tips on good practice and pitfalls to avoid.

In fact most of the information is included to enable any competent person to engage this profession and start manufacturing saleable products within a few weeks.

It may become obviously quickly, but this book has been written by a Fabricator and with the assistance of similar folk, we are not writers and this is our first attempt!

I had to put this in, don't be like this chap, btw he did actually exist,

The Metal Fabricator's dilemma

Local Council business inspector arrives at 'Too Cheap Welding Services'

Council Chap: 'Good Morning Sir, I'm just conducting a survey on local business and its employee's. Good you tell me how many people you employ and what do they do? Also, how much do you pay these guys?

Gaffer: Well, old Tom over there, he works the Guillotine and Pressbrake like the other guys, he works 37 ½ hours a week, gets sick pay, plus 28 days holiday and 8 statutory paid holiday. He uses the works truck on a Sunday for car booting and I let him store his caravan in the yard free of charge. He gets £350.00 per week. Then there's Mick, he's the Welder, moans constantly about requesting that the company buys better extraction, ventilation and the latest 300+ quid welding masks, he needs lots of breaks to get fresh air outside where he feeds his 60 a day fag addiction. Very often he brings his garden rubbish and fills my skip. He Gets £400.00 per week. There's also Harry, loses lots of time because he struggles to get up in the morning, has to pick his kids up and finishes early on a Friday, his Mother in law has died four times so has lots of time off for funerals and every time he go's out for a beer in the evening, strangely, his car won't start the next day, and we don't see him. He gets £300.00 a week.

Council Chap: Ok, anyone else?

Gaffer: Well there 'Dave the Halfwit', he works 18 hours a day, can work all the machines, he works most weekends, spends a lot of his time putting jobs right, hardly takes a break, loans the company money in hard times, which he never gets back, puts the other employees welfare before his own, oh, but on occasion he gets to sleep with my wife. He gets £90.00 per week, if he's lucky.

Council Chap: Oh, really, I must say Mr Gaffer I'm disgusted to hear about this poor fellow, where can I find 'Dave the Halfwit?'

Gaffer: You're talking to him.

Theres a moral in there somewhere, can you see it?

Consumables? What will I need?

CHAPTER 5..................... *Equipment*

What do need in my toolbox?

What Power tools do I need?

What plant and equipment do I need?

Workshop Layout

CHAPTER 6...................*Working Safely*

P.P.E

Site Safety

Health Authorities

First Aid

CHAPTER 7.................. *Sample Paperwork.*

Quotations

Purchasing

Delivery Note

How to raise an invoice

Statements

CHAPTER 1.....................*Where do I Start?*

Ok, so you've decided you want to have a go at manufacturing Ornamental Gates. Do you want to make them for a special project, maybe at your own house or business premises? Or as a part time hobby, maybe sell them from time to time, or do you want to take the plunge and go full steam ahead and manufacture full time?

Do I really want to run my own business?

Ok, let's look at *some* of the Pro's and Con's.

Pro's:

•You get to work for yourself, you can be proud of your achievements!

•You make the decisions

•You decide what hours you work

•If it works out well, you reap the rewards

•In years to come, the business may be worth a lot of money and could form a basis for your retirement pension needs.

•You are the master of your own destiny, you should aim to take out what you put in

Cons:

•Working for yourself can be lonely, swift and accurate decision making are a must, it can be difficult with not having anyone to 'bounce' ideas off

•Hours can be long, lots of effort needed to get sales going in the early days, if you don't like hard work, it's not for you.

•Sometimes, you have to draw that line and realise that there are 'no friends in business', problems can result when dealing with friends or family if they fail to pay or are not happy with the goods.

•If it all goes belly up – depending on what you've agreed to risk, it could be financially disastrous

Self-Employment or a Hobbyist?

For the purpose of this book, I'm assuming that yes; you want to dip your toe and have ago on a self employed basis! Hopefully, this book will help you to understand the pitfalls and risks but also the potential of high earnings and an exiting opportunity to join the ranks as a small business owner. As a hobbyist you should have enough information after to reading this book to at least make an attempt, if you have any basic engineering or Welding experience then you should have no problems. You have to notify the Inland Revenue of your intention to go self employed, as the law stands you do not legally need an accountant and do not have to submit accounts. A good accountant is good practice though and should save you more money over a year than he/she costs you.

Will it be easy?

No, it will be hard work, but rewarding in the long run, even if you decide you want to do something else with your career, you will have gained extra and worthwhile skills for life.

First Rule – In it to win it.

You need to make a profit and not be a busy fool, be selfish, think of yourself, your goals, and your long term plan. If you end up employing others or expanding or buying new plant, these actions must add value and contribute to a healthy return for **your** business model! In the early days, and as the old adage goes, you may need to 'throw a Sprat to catch a Mackerel' in other words quote a job cheaply to get the job, impress the customer and expect repeat business, the downfall with this is next time he may still go out for the cheapest price, whilst being the cheapest isn't a bad thing in itself you need to remain profitable or effectively you are **'buying work'.**

Employing Others?

It may be inevitable in the long term, but initially try and keep your costs low, maybe use subcontractors or Temp workers to fill in during the busier periods if you cannot cope, alternatively, you could subcontract work out. If you employ a person or if you decide to start up as a 'ltd' company you need, by law, Employers Liability Insurance. You would also need Public Liability insurance if you decided to fit the Gates and Railings at peoples or Business premises.

How much do I already need to know about Metal Fabrication?

Everything to do with Fabrication can be taught, lets face it, if your smart enough to buy a book for advice because you're not 100% sure , then your already on that learning curve. The skills required are mainly among those that any DIY'er would posses, measuring, cutting, cleaning, drilling, etc... For those that haven't any Welding experience this will be the hardest skill to master, but can be self taught. There is a wealth of info on the web on this subject, and I will cover this in a later chapter with advice on the type of equipment etc....I am also assuming that a lot of you guys are already proficient welders or Fabricators, therefore may just require some pointers in the direction of starting a business or the Manufacture of Gates and railings.

Why do some business's flourish and some fail?

Location? Could be the wrong type of business for the area, Break ins? Lots of small industrial units are broken into, businesses have gone bust for this reason. A business selling Ornamental Ironwork, situated on a main road in an affluent part of town would fare better than a small unit in the centre of a run down estate where unemployment is high and money is scarce.

Costs? Move into an expensive 3000 sq ft unit on your own and expect to be able to keep it productive, efficient and busy? Or do you share a unit, maybe just rent a bench space until you start building relationships and see the business grow.

Salary? Ok, you've walked out of your 25k a year job as a Production Manager or Welder and straight away start taking the same amount from your new business, because its what your used to and you need it don't you? The successful guy budgets, covers his overheads and takes a percentage of what's left, it will take time to build up to a decent , regular wage.

Lack of work? The biggest threat is not keeping busy, time is money. It is difficult to work on your own and go out and bring in new work. In this trade, without a doubt the best advertising media is 'word of mouth', keep customers happy. Stick to the original costs, if you're fitting the gates then sweep up afterwards, offer to get rid of any debris, etc... You will find that if you're constantly busy and can't keep up, and also this will happen, you maybe could afford to prices up a little. Diversify is you need to, Production welding of small components is another avenue. Your workload will always consist of 'Peaks and Troughs', plan your workload and cashflow accordingly.

Competition?

You may chose an area to apply your marketing strategy, maybe an area where there isn't an Ornamental Gate maker within 10 miles, this maybe is because in this area these items don't sell? Could be run down estates with no cash or new developments with no permissions for railings or gates... on the other hand the adjacent area has 4 manufacturers within 3 mile radius – there you go, you've found an area with a **demand**, sometimes this demand will not be great enough to sustain multiple Gate Makers, find out, you have to visit these guys, Investigate, ask questions! This is front line Market Research, and a necessary evil, if you ask the guy how long to make a set of drive gates and he says 8 weeks then he's busy, if he says 2 days, then he Isn't! You will find you take a little bit of work from each of the other guys and in the interests of healthy competition you need to see where and how they advertise, do not just try and undercut, but try and find reasons why you can charge more for your product! Quality, guarantee, Fitting discounts all help. In dealing with the public personality counts, these people need to *like* you, before they will recommend you.

Am I selling a service or Product?

This is simple but important, you are selling both. The local builder or the householder will want the Gates fitted, so you're selling the Product and the service of fitting them. The internet sales will be supply only, these are two very different markets, your website or eBay shop will probably have to be more competitive and the range should be standardised as much as possible, your local work will be more bespoke, fancy (probably using better materials) and with a healthier margin. A product (such as an EBay sale) is a quick cash turnover, a sale including measuring , designing , manufacturing then fitting could affect your cashflow should things go wrong and payment gets delayed.

How to raise an invoice

This is done in a very similar format to a Purchase Order. Customer and Your details, Description of Items, Date invoice raised and terms. **A Pro-Forma** invoice is used when you collect the full monies up front. The invoice must include VAT if applicable and state the VAT number on the paperwork. See the example in the later chapter for guidance.

CHAPTER 2……………….Where will I be working from?

Ok, we all have heard of people in 'cottage industries' working from home or in their shed, is it realistic to think its possible to work from home, making a products such as these?

Yes, it can be done, not easily, but I'll explain what I did years ago.

I had a fulltime job as a Fabrication Manager and decided to have got at making Gates and Railings in my spare time and weekends. I had a workshop at the bottom of my garden, it was a steel type, sort of like corrugated cladding and was quite large at 20 foot x 12 foot, in fact it was two purpose made steel sheds that I fixed together. The floor was slabbed, important as a wooden shed may ignite with all the welding, cutting and grinding heat and resultant sparks. You are, however, limited to what equipment you can have do to the domestic UK supply only being 240v, so three phase is a no I looked at expensive three phase converters but these seemed more practical for motors rather than welding equipment. I did have a single phase Arc Welder, it was an ESAB invertor welder which served me in the workshop and onsite well for several months. Noise was a problem and late night working was out because of disturbance to the neighbours, the saw, an abrasive wheel cut off saw was very noisy, even my wife complained. I discovered arc welding was too slow and expensive to efficiently manufacture Gates and railings so I looked for a single phase Mig welder, bear in mind I had, in my shed the power equivalent of a cooker/shower outlet of 30 amps, by the time this tricked down my armoured cable into my shed fusebox it was probably only 20 amps, I'm not an electrician but it works something like that! I was limited to what I could buy, I managed to get a second hand 220 Amp welder, good for 8mm thick, which was good enough for now, but the duty cycle was very low, which means it keeps cutting out and has to cool down, which was not as much as a problem as you would think, welding quantity on a Gate is very low, Cutting and drilling metal, squaring and tacking the job together takes the time. Some of the later jobs I took on involving heavier section metal, RSJ's etc... I couldn't have done. Thus, I would definitely need a 350Amp 3 Phase machine.

Spray painting the finished gates in the garden? Forget it, neighbours will be out in force complaining! Hand painting takes too long, sub-contract Powder coating is the only option.

Add to this the problem of informing your mortgage company or landlord that you will be conducting a business from home and after informing the local business rates department you could lose any advantages that you would think working from home has to offer.

People aren't going to drive past and notice your business? That's a big negative. Steel stockholders deliver Steel flat in 6metre lengths and Box section in 7.5metre lengths, can you handle these?

Conclusion: it worked for me when I did it on a small part time basis, but very soon, just a few weeks later I knew I needed to bite the bullet and get an Industrial Unit, however, I decided, it would be best to retain my office set up in the bedroom, I would work in the factory in the day and do my paperwork at night at home, this worked well, but the hours were long.

An industrial unit is, by far the best option, try and choose one near a main road (or just off a main road you can always have signage pointing the way) with passing traffic. The ability to accept a sizeable truck when you get deliveries would be useful. A clear purpose made sign is a must, don't go down the hand painted sign route, people want neatness, tatty sign= tatty gates! Generally, the smaller the unit the higher the price per square foot, negotiate, look for rent free trial periods, and look for shared units. For one person, a starter unit of 500 sq ft should be fine, this will allow a saw area, bench space, drill table, Welding Bench, etc... My first unit was 500sqft, eventually we had two persons operating with no problems.

So for costing purposes you need to find an industrial property landlord, negotiate a rent free period or even a no deposit scheme would be of help. Find out the cost per sq/ft and the amount of sq/ft available. Try and ensure the landlord included building insurance and maintenance costs (these sometimes show up as an expensive extra).

You also need to find out if the electricity bill is paid to the landlord (some have their own meters and charge a little bit more per unit) and similar with the water, is there a water meter?

You can do a quick sum. Landlord has a nice clean unit 475 sq/ft @4.85 per sq ft. 475 x 4.85 = 2303.75 Inc vat. (If payable) divide by 12 months and the rent is £192.00 per month.

Assuming (actually a must) that the unit has three phase fitted, you may need an electrician to do additional work such as fitting welding sockets, etc...

Now this is very important, people underestimate commercial business rates they can be very high, check with the local council for the rateable value of the property and how much you will likely be charged, normally there is a discount for sole traders or businesses with just one person working there. Some landlords may include rates, and recharge with the rent.

There is really no need to go for long leases, stick to and insist on monthly leases, this means should the worst happen and you can't carry on trading you will only be liable for one months rent. Whereas, if you signed a three year lease you would be held accountable for the remainder of the lease, personal bankruptcy has in the past, been the only way to avoid paying off a lease when your sole trading company goes under. A long lease should only be considered if you chose the ltd company path.

CHAPTER 3....................*The Administration side of it.*

Marketing, Advertising and Selling

There are lots of avenue to get in to potential customers, firstly, lets look at who some of your potential customers, and best methods of sales catchment are:

- local Homeowners -Local Press, Passing traffic & Flyers

- Builders- E-Mail shots, flyers.

- Property Developers-E-Mail shots, flyers.

- Industrial Buildings (think Security Grilles) E-Mail shots, flyers.

- Local councils, Schools. - Look for advertised tenders

- Website sales- Aim to get good Google results. There are many options available for free websites that are very easy to build, for a small charge these website companies can greatly enhance your web search results.

- EBay Sales – whether you sell directly on the eBay site or customers phone or email you their sale, there's no denying the power of EBay as a sales tool. You will get sales, there is lots of competition don't aim for being the cheapest aim at looking the best and most professional, push the Made in Britain theme, there are lots of inferior Eastern European and Chinese Gates and railings out there, they are not bespoke, made to measure and usually not fully welded.

An advert in Yellow Pages or Yell.com will almost certainly yield a response; however, you have to wait for the next print issue and should be viewed as a long term strategy.

Costing's and Quotations

Before you attempt a quotation you need to evaluate and determine what your fixed costs are, there are lots of ways of doing this but for a Sole trader (when I keep saying sole trader this could also mean partnership) business we can keep this simple.

Let's put some basic fixed and variable monthly costs down. Lets assume you don't need a phone line or internet, I say this because the internet you probably have at home already and most sole traders have a mobile anyway, complete with internet/email.

Lets assume you loaned your company £2000.00 to help get started, the company will pay it back without interest at £166.66 per month over 12 months. If you borrow from bank then just add the interest.

Initially, I would assume that subcontract local transport would be used for deliveries, for web based delivery then a parcel carrier.

Before you start up you should ensure you have two months salary in reserve, this will help until the money starts coming in, I've allowed for a salary in the costs of £1500 per month, for costing reasons.

Rent   £250.00 per month (Inc water)

Rates   £124.00 per month

Electricity £85.00 per Month

Mobile Phone. £30.00 per Month

Insurance: Public and Employee Liabilty £147.00 per month

Loan £166.66

Salary £1500.00 x 1 person

 Total: £2302.00 per month direct costs.

£2302.00 costs per month equates to about £110.00 per day costs,assuming that there are 21 working days in a month.

1 day divided by 8 hours per working day = £13.07 per hour cost.

This cost £13.07 assumes that you are 100% efficient, 85% or there abouts is more likely, let's give us some scope for error and say we need £15.50 per hour to break even.

Lets look at an example.

Mrs Smith requires bow top gates , supply only to fit a gap of 3000mm wide x 1550 high, powder coated black.Customer collect.

Labour: 2 1/2 hours cutting and forming, 2 hour drilling, 2 hours tacking up square, 1 hour welding,1/2 hour linishing and dressing welds.

8 hours labor x 15.50 per hour=£124.00 add £35% margin= **£167.40**

Materials:

1 ½ length of 30 x 30 x 2mm box ERW £12.00

2 lengths 30 x 8 flat   £16.00

5 lengths of 1/2" solid dia bar £4.00 ea £20.00

Latch x 1 @1.90

Hinges x 4 @ 1.00 each £4.00

Spearheads x 15 @ 0.35 each £5.25

100 dia Circles x 20 @0.40 each £8.00

*Total £67.15 approx add £25% margin for materials.***£83.94**

Powder Coating: £27.00 add £25% margin. **£33.75**

Total selling Price £285.09

This should be your minimum selling price, you can add extra margin if you think the job can stand it, if dealing with builders or developers who pay on 30 or 60 day terms (thats end of current month + 30 or 50 days) you may wish to add a factor on to accomadate the extended payment. I used to add on 15% and discount it if early (7day) payment was agreed.

If you register for vat, which, in itself is a very easy process, you will be more expensive when selling to the public,i.e. Homeowners and such like, you will have to charge VAT, currently @20% . However, all of your suppliers will be charging you 20% Vat on materials and services which you be able to claim back at each quarter. Without being VAT registered you will be uncompetitive. There is a threshold that you will by law have to charge for VAT, but it is good practice to register anyway.

Its important to note that the above costs assume that only one person is employed, this person then has to cover the direct overhead costs in his chargeable hourly rate. It is right to assume that when the business is able to take on an employee the overhead recovry costs for the rent and rates, etc,.. are shared. This brings the chargeable hour rate down and make the business more competive, assuming your efficiency is maintained with enough work for two persons. Your new Chargeable labour rate, assuming the new guy is directly employed and subject to employers NI contributuions would drop to between £12-13 per hour.

Fitting should be charged at the labour charge per hour+ Transport+ Materials I.e postfix or cement + 20%.This will give you scope to subcontract out during the busy periods.

Note:

Try to keep your material costs below or around 1/3 rd of the selling price.

For local work, especially bespoke ask for a deposit of 30%, with the balance on delivery/installation.Ebay/Website should have abuy it now button, be wary of Paypal and Ebay charges , add these costs to the sales price where possible, you will quickly see you Ebay charges rising the a couple of hundred quid per month, right off your bottom line!

Your quotation should be clear and concise, it should show a breakdown in costs, I,e item cost, delivery charge, vat etc...

lead times: try and stick to deadlines, customers will claim they had six guys on site who travelled four hundred miles to fit the gates that never arrived as promised! If possible provide a sketch or picture to show the customer what he is getting, if its a bespoke job and someone agreed over the phone to make a gate 6ft high x 5 ft wide, and it was made 5ft high and 6ft wide then you may have aproblem, send a sketch for approval,ideally before work commences.

Purchasing

When times are hard and prices are tight the profit (or not) on a job may be in the buying. Always get several quotes, always negotiate, sometimes 'oddsize' steel may be available, maybe an imperial size such as 38mm wide flat instead of 40mm, if you can get agood price then its worthwhile considering.

There are lots of companies selling railheads and other ancillaries, try and buy in bulk, use standard railheads and try and use common stocked steel sizes.

Receipts and Invoicing

Receipts may be required for deposits or cash payments, a small duplicated book with a letterstamp with your details on would be good enough.

Invoicing can be done easily by yourself, ensure invoicing is done at least weekly as it will help to stay in control of finances, there are many standard invoicing forms on the internet, which are freely available.

The Invoice should have the company name on, address and also the Soletraders name such as *Bill Bloggs 'trading as' 'Gates and railing services'*.

It should also have a VAT reg number if you have one and a breakdown of the amount of vat charged. An example invoice will be shown in a later chapter.

Insurance

You need by law Employee Liabilty insurance if you take on an employee,or decide to create a LTD company.

if your premises are also open to the public or indeed you receive visitors then you need Public and Employers Liability which are normally offered as a package.

You should also check and make sure you are covered whilst fitting and ensure they know there could be welding or 'hotworks' onsite.

CHAPTER 4.................What *Materials do I need ?*

What Metal should I buy or Stock?

Ok, a common misconception is 'Wrought Iron' is seldom used, lots of manufacturers adverstise 'Wrought Ironwork' , this more a Blacksmithing job were items are hand forged, typical large fancy gates that frequent an old church or the gates of Buckingham palace are examples of this, for our purposes we use Mild Steel flat and 'Box' Section. This is the easisest metal to work, form and weld and the most readily available.Using hollow swction will keep the weight down, so using 5/8" Hollow tube for the vertical infill bars will be lighter and cheaper than using 16mm solid round – the finished job will look the same, the weight reduction will help when fitting and minimise deflection.However, the exception to this rule is that some customers will want the best and heaviest section, complete with expensive artisc infill panels, this is fine and great work to get, but when choosing the right materials for the job it is imperative to get this right.

Steel Stock, Typically you would order similar items to this list below:

30 x 30 x 2mm ERW square tube, 6m length- Usage- Outer Frame

40x 40 x 3 SHS -7.5 metre lengths- Usage Posts and Outer Frame

50 x 50 x 3 SHS-7.5 metre lengths- Usage -Posts

80 x80 x 4 SHS-7.5 metre lengths-Usage -Posts

100 x 100 x 5 SHS-7.5 metre lengths-Usage -Posts

30 x 6 Flat- Frame and Horizontal members- 6m length

30 x 8 Flat - Frame and Horizontal members- 6m length

40 x 8 Flat - Frame and Horizontal members- 6m length

40 x 10 flat - Frame and Horizontal members- 6m length

12mm round bar – Vertical Infills 6metre length

16mm round Bar Vertical Infills 6metre length

5/8" x 2mm Tube for Vertical infills x 6 metre length

12mm Square bar – Vertical Infills 6metre length

16mm square Bar - Vertical Infills 6metre length

Other items you will need can be made or bought in, these are items such as hinges , latches, dropbolt, scrolls. Due to comparative low cost of these items and the fact they look better than most of the handmade efforts I have seen I would advise that you purchase these, unless of course you enjoy making them!

Spearheads or Finials will be used in great quantity, you should try to minimise the amount of different designs as some look quite similar, try and offer the customer what you have in stock. Some infill panels may be bought in 'cast Iron' , be sure to only purchase Mallaeble 'weldable' items, examples are below:

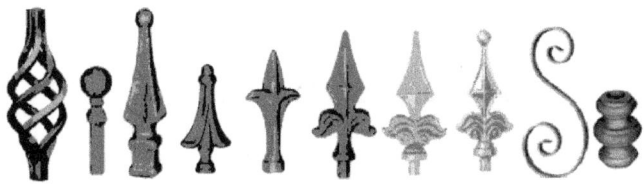

Consumables? What will I need?

Ok, the last thing you want to do is be working flat out , getting a rush job ready for the Powder Coaters the next morning, then you find you've run out of Mig Wire, or Welding gas or Grinding discs, and so on. Make yourself a rule, one day you go out and pick up consumables, assuming that is you can't get them delivered. Working alone is hard, you need to be motivated it's is easy to have a break from the monotony and pop out for a tin of Paint, it takes you half an hour, thats half an hours production time you've cost yourself. That £10.00 tin of paint has cost you nearer 20 quid.

So regular use consumables would include:

Welding gas – You need to register and pay monthly rental on the bottles, easy enough to set up though.

Mig Wire

Arc Rods

Torch tips etc...

Anti Spatter spray

Gloves

Safety Glasses

Welding headsheild Lenses

Linishing and Grinding discs

Drill Bits

Marking Out Chalk/Markers

Packaging goods- Tape ,Bubble wrap.

Etc...

Whilst planning your new business or Plans to start making these products start to purchase these items of consumables, you will be glad you did.

CHAPTER 5.................. *Equipment*

What do need in my toolbox?

Tape Measures , a 3 metre for workshop use and a 5 Metre for site surveys.

Spirit level (and possibly laser level)

6" Steel rule, !2" Steel Rule and 1Metre long Rule.

4" and 6 " Engineers Squares

Comination Square/Slide Rule

Scriber

Centre Punch

Trammel Heads

Dividers

Ball Pein hammer

Lump Hammer

Sledgehammer (For Site Installation)

Large Prybar/Crowbar (For Site Installation)

Vernier

'G' Clamps 2",4",6".

Sash Clamps

110v and 240V extension leads

What (110Volt) Powerered tools do I need?

4 1/2" Angle Grinder- For linishing and grinding -go for a branded one with the highest wattage rating. (110Volt)

Hammer Drill (110Volt)

Cordless Hammer Drill

Pistol Drill (110Volt)

110 Volt Transformer

What Plant and Equipment do I need?

Workbench- Tack the gate up un a workbench

Trestles -Once tacked, weld the gate on the trestles, this will make life easier , the Weld Spatter wil tend to drop on the floor rather than stick to the bench

Pillar drill –a good secondhand one (Meddings,Fobco,Starret etc...) would be a better buy than cheap new one. The chuck must be the 16mm (5/8")type and the motor suffiently poweful enough to drill a 16mm hole in 12mm thick mild steel.

consider, give these a wide berth if you can. The heavier industrial ones are superb, but expensive. Excellent if you can afford one.Metal Saw – You have several options here, you can buy a mechanical hacksaw, which you can leave to cut the steel, it will go back to start on completion of the cut.

Then there is the horizontal bandsaw,shown below left, the cheaper Diy or occasional user models are not robustly made enough to consider, give these a wide berth if you can. The heavier industrial ones are superb, but expensive. Excellent if you can afford one.Metal Saw – You have several options here, you can buy a mechanical hacksaw, which you can leave to cut the steel, it will go back to start on completion of the cut.

More

common are the 14" (355mm) Abrasive cut-off saws (shown above right), these are really not a saw but a large 14" Cutting disc, they are cheap to buy (around £200 new) and easy to use, you may have to deburr the edges after cutting and the amount of grinding dust will make your workshop very dirty and is very unpleasant when it gets in the air.

Industrial Circular steel saws are very good, quite, quick and not too messy, these can be expensive but occasionally come up second hand.Recently, other cheaper saws have entered the market, one in particular is the TCT 14" Evolution Rage 2, these are about £200 new and come with a blade, to replace the blade cost around £60.00, expensive considering you only get between 500 and 1000 cuts, the quality of the cut is cut though, and like the Abrasive Wheel Chopsaws will cut up to and slightly more than 100 x 100 box section.

A set of rollers or Material support trestles will assist greatly when pulling the material onto the saw table.

Bench Grinder – This will prove useful for sharpening drill bits and keeping your Scriber in tip top shape.

Compressor – A must if are considering applying primer or Painting your gates yourself, choose one with aminimum 100 litre tank and at least a 3 HP motor,Spray guns can be bought cheaply and Air tools such as drills, grinders and linishers can be used if required.

Bending rollers-Used to create rings , a non-essential piece of equipment but has its uses.

Ironworker- Ok maybe not on day one, but these machines punch holes, guillotine flat bar, cut Angle iron, notch and shear plate, workhorses that yu can live without but life is simpler with one!

Arc (Stick)Welder- A must if you are contemplating sitework, sometimes you may have to weld hinges to existing steel posts, a ten minute job with the Arc Welder, it will also prove invaluable for quickly taking up jobs in the factory where dragging a Mig Welding across the factory proves tricky. Try and get an Invertor Arc welder (usually these have a Tig weld option as well), these can weld some pretty thick plate even when used with a long extension lead, duty cycles are good and they are much easier and relliable to use than the other non-invertor portable Arc welders.

Mig welder- You wont get far without one of these. Aim for a 3 phase set,around 300+ amps, complete with an MB36 EuroTorch (the Smaller mb25 torch will not handle the amps on the thicker material). Look for secondhand Murex/BOC/Thermal Arc/Miller/ESAB/Lincoln/Butters etc... all very good makes and repairable when things go wrong, ask your local Welding machine supply about second hand machines with a Warranty. If you decided to buy new you could pay £000's, a new clarke machine at a grand may not be as good a buy as a Second hand Murex From Ebay. If you decided to go for new , try and get extended warranties and be sure of spares and back up when required.The welding wire is available in different thicknesses, however 1.00mm should be fine. This comes in reels that are 15kg in size, this is the standard for Mild Steel Mig wire.

A Suitably sized Mig Welder is shown Below. Ideal for Gate Work.

Vehicle – I made do without a van for several years , I used external Transport and sometimes even external fitters, desirable but not essential.Eventually I made a purpose made trailer for carrying gates, it saved me a fortune actually and I didn't have the burden of running a van when my car would suffice.

Workshop Layout

You need to design your workshop how you want it, however I would highlight these points:

•Make yourself a steel rack or cantilever brackets of the wall, these should be position nearest the entrance where your steel would be delivered, it makes easier to put into stock, keeps it off the floor , £/sq ft is expensive , keep your floorspace clear

•Put the saw adjacent to the steel rack, obviously.

•Put the welder at the furthest corner, put weld screens around it, to minimise the possibility of 'Weld flashing' a co-worker or even worse a customer! You may need to luck at fume extraction, especially if you weld Galvanised steel.

If spraying, follow guidelines by paint manufacturer and equipment seller, Keep paint and thinners in a locked steel cupboard, check periodically for leaks – keep away from Welders or any sources of ignition.

•Scrap/rubbish bin, *clean as you go.*

•If you have an office great, if not have a clean table near the entrance with business cards. Literature and any other details customers may need.Have a folder with pictures of previous jobs or pictures of accessories.

•Sample Wall, have a few designs and samples fixed to the wall, let people have confidence in the quality.

•Fire extinguisher/ Fire Bucket or Fire Blanket are strongly advised. Ditto for First Aid kit.

CHAPTER 6.................*Working Safely*

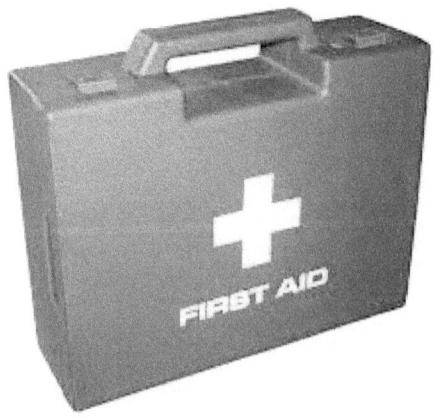

This topic is perhaps the most important and would benefit from extended reading and just browsing through the HSE website wil give you lots of advice and preventative measures that you should consider putting into place.

P.P.E

Personal Protective Equipment, in your case may mean the following items: safety galsses, Safety Visor, Welding Mask c/w extraction module, Filtered or Air Fed mask for spraying,Gloves for handling sharp items, Welding Gauntlets, Fireproof overalls, Safety Shoes/Boots and if your working On A Building Site Hi Vis Vest and Hard Hat as well. These are a must, no excuses. Follow the instructions for any hazardous substances you may be using and also consider P.P.E eqipment if you have visitors wandering around your workshop.

Site Safety

If you do win a job where you have to fit Grilles,Railings, Gates etc... on a building site or other Industrial/Commercial premises they are several other things to consider. You may have to attend a 'Site Induction' this is where the person onsite who is responsible for Health and Safety will give you an outline of the site and explain potential hazards to you. He may also show you were First Aiders or First Aid stations are situation and what to in the event of an emergency.Exit routes will be explained and you should listen to everything you are told, if the guy says no smoking , then don't let your guys smoke! These inductions will differ from site to site. You may have to be certified prior to going on site , this may be by using the CSCS registarion systemwhere you need to attend a course and sit an exam at the end, other pprojects have also been introduced, Passport To Safety is another, but CSCS which has different levels, Labourer, tradesman, Supervisor, etc... is the most requested.

You may also have to provide a Method Statement, where you need to state information such as, the name of your customer, The name of your company,The names of your Site Workers, Where on Site you are working, Contact details, What you will be doing on site, the processes involved,what equipment you will be using, Job Start and Completion date etc....

If they ask for a Method statement then they almost certainly ask for a Risk Assessment, where you have to asses the likelihood of an accident happening, the level of severity should an accident happen, and the steps you are taking to risk the chance of the accident happening, Again the HSE website is invaluable here.

Health Authorities

Shops and commercial properties are normally overseen by the local councils own Health and Safety officials, factories are looked after by The H.S.E, The Health and Safety executive. You can search there website to find out which one is responsible for your area.

However, local authorities will also have environment al officers so if your Welding Extractor or Painting Booth is emitting pollution over the acceptable level you could expect a visit, they do monitor Air Pollution in the atmosphere and can easily tell where the source is coming from.

First Aid

Ideally you should be qualified in First Aid and know the law regarding first aid at work, this may prove problematic if you work alone, if you are working alone let someone know of our whereabouts you could be injured and unable to call for help.

Make sure you always have access to a phone. A first Aid kit is a must.

CHAPTER 7................ *Sample Paperwork.*

Quotations

Ensure sure you Show all the chargeable costs, these could include Manufacture, Powder Coating, Delivery, VAT. As well as discount for early payments or agreed deposits. Try and leave no scope for the customer to query costs, quotations must be concise and also state disclaimers where necessary, I.e. 'we will not be responsible for removing and disposing of the old fencing.' this sounds obvious but all 'i''s must be dotted and 'T's must be crossed!

Purchasing

This Document must show the following details:

The Date, Your Name, Address, Contact no, Email, Fax number if you have one.

The Suppliers Name, Address, Contact no, Email etc...

Description of items bought, be accurate or you will end up with 100 off 4" Cutting discs that won't fit your 4" Grinder! , get the price right, include agreed discounts. Include the Date you need the items for, this is commonly known as the *Lead Time*. State on your order whether or not carriage is included or list as a separate charge. If you have agreed terms state these too. If you have agreed terms such as End of Month + 30 days, try not to place and order on the 30th, place it on the 1st, this will give you more time to pay and help cash flow. Don't pay early unless a discount is offered.

Delivery Note

This Document must show the following details:

The Date, Your Name, Address, Contact no, Email, Fax number if you have one.

The Customers Name, Address, Contact no, Email etc... The date Shipped, an exact description of the goods sent, you must note that any damages or missing/incorrect items must be notified immediately.

Send two copies one for the customer to keep, and one to sign and return (in the case of couriers they will collect a signature using digital means).

Invoicing

Invoicing will be the most important document you issue.Try and invoice as soon as the jobs go out, stay on top of the invoices , you will get customers who have mislaid an invoice, not received it or are just slow to pay.If the invoice is not paid within its due date then wait for a week or so and contact the customer.Post the invoice to the customer but if possible deliver by hand with the goods.

Statement

A statement is sent to each customer on a monthly basis, the statement will show the invoice number, the date the Invoice was raised and the customers Balance of unpaid Invoices to date.

ABC Welding Services

Unit 6, The Industrial Estate, Anytown, AT1 0AB

Tel: 0777888999 Email ABCWeld26758@aol.com

INVOICE

Customer Details: B Smith Builders **Customer O/No**: SF/3455

Invoice Number: 012343   **Date:** 22/06/12

To Supply 1 x set Bow Top Gates to suit

opening 3560 w x 1550 h, Powder

Coated Black with Gold Highlights	£490.00
600High x 1800w Railings to suit 10 sets.	£1550.00
@ £155.00 each	
Delivery	F.O.C
Installation 1 day x 2 persons	£425.00

Terms: ***To be Paid Within 7 Days***

Total £2465.00

Vat @20% £493.00

Grand Total £2958.00

Proprietor: Mr Bill Bloggs …..VAT Registration 34553765

ABC Welding Services

Unit 6, The Industrial Estate, Anytown, AT1 0AB Tel: 0777888999

Email ABCWeld26758@aol.com

Purchase Order

Supplier Details: B Smith Steel Supply **Our O/No**:
SW/220612/02 **Date:** 22/06/12

DESCRIPTION	QTY RED'D	DUE DATE	TOTAL COST+VAT
30X30X2 ERW @ £5.00 EACH	12	24/06/12	£60.00
30 X 8 FLAT@ £5.50 EACH	5	24/06/12	£27.50
12MM ROUND @ £4.00 EACH	50	24/06/12	£200.00
50X50X3 SHS	1	24/06/12	£22.00
40X40X3 SHS	1	24/06/12	£16.00
80X80X3RHS	1	24/06/12	£49.00
50X6 FLAT	1	24/06/12	£8.00

TOTAL COST £

382.50

ABC Welding Services

Unit 6, The Industrial Estate, Anytown, AT1 0AB

Tel: 0777888999 Email ABCWeld26758@aol.com

Delivery Note

Customer Details: B Smith Builders **Customer O/No**: SF/3455

Invoice Number: 012343 **Date:** 22/06/12

To Supply 1 x set Bow Top Gates to suit

opening 3560 w x 1550 h, Powder

Coated Black with Gold Highlights

600High x 1800w Railings to suit 10 sets.

Please sign for receipt of goods above in good condition:

Customer Name (Print)...

Customer Signature...

Delivery date: 22/06/12

Please report any damage or missing items within 7 days*

Proprietor: Mr Bill Bloggs …..VAT Registration 34553765

ABC Welding Services

Unit 6, The Industrial Estate, Anytown, AT1 0AB

Tel: 0777888999 Email ABCWeld26758@aol.com

Statement of Account

Customer Details: B Smith Builders

Date: 30/06/12

Invoice Number	Date of Invoice	Invoice Value	Comments
12343	22/06/12	£2,958.00	
12312	01/06/12	£256.00	
12301	01/06/12	£306.00	

Invoice Number	Date of Invoice	Invoice Value	Comments
12205	25/05/12	£3,566.00	
12106	12/05/12	£122.00	
0-30days-current	31-60 days	61-90 days	91-120 days
£3,520.00	£3,688.00	-------------	----------------

Proprietor: Mr Bill Bloggs …..VAT Registration 34553765

ABC Welding Services

Unit 6, The Industrial Estate, Anytown, AT1 0AB

Tel: 0777888999 Email ABCWeld26758@aol.com

QUOTATION

Mr D Jones, 22 Rosebud Gardens,Anytown AT1 2AB

To Supply 1 x set Bow Top Gates to suit

opening 3560 w x 1550 h, Powder

Coated Black with Gold Highlights	£490.00
600High x 1800w Railings to suit 10 sets.	£1550.00
@ £155.00 each	
Delivery	F.O.C
Installation 1 day x 2 persons	£425.00

Lead Time: 4 weeks from date of Order

 Many thanks for giving us the opportunity to quote, please don't hesitate in contacting the underside should you require more information.

Yours Sincerely,

Bill Bloggs

ABC Welding and Services

Proprietor: Mr Bill Bloggs …..VAT Registration 34553765

CHAPTER 8..................*Manufacturing gates – How to?*

Measuring up & Design

Both the measure up and design can be done by yourself (assuming local), or by your your client, if the latter the client assumes responsibility.

When Measuring up for Gates measure the overall width of the area to 'fill'.This area will include posts if required (or the gates may be fixed to existing brick pillars perhaps?) , hinges, gates and the centre gap, normally 10mm or so. Measure the top and the bottom widths. The adjustable hinges will compensate if there is a slight (say, 50 to 75mm discrepency).

The Height from the ground to the highest point of the gate should also be measured. You should assume to have around a 50mm gap from the ground to the bottom of the gates, this may vary if the ground is on an incline or slopes upwards.

The ground may also slope from left to right, if its more than about 50mm over a width of 3 metres or so it would be best to taper the bottom of the gates to suit. A long straight edge c/w a spirit lever will give you the fall size, simply replicate this on the gates, there is more detail on this in the final chapter.

The next step after measuring up is the design, the customer may want large heavy gates but insists they are fitted to his 100 year old stone 'falling apart' wall! Stand your ground and advise smaller/ lighter gates, or advise posts will be required too. You will need a catalogue of designs or pictures, let the customer chose and quote accordingly, feel free to Use the designs at the back of this book to help you.

Remember: Measure twice, Make ONCE!!

Cutting Lists/Templates

Roughly sketch your Gate opening out, and the sketch the Gate frame and shape into the opening. Lets say the gate opening is 3540 wide x 1500 high in the middle and 1300 high at the side brick pillars.

The Hinge Brackets and Adjustable Hinge Bolt have an allowance of 100mm each side (they have adjustment of around +/- 50mm both sides,see pic below),

so thats 200mm, add to this the 10 mm gap in the middle which makes 210mm. Therefore 3540 – 210 = 3330mm of gate, supplied in two sections = 3330 / 2 = 1665mm each leaf overall wide.

The gap beneath the gates is 60mm and the spearheads are 120mm high,so the centre gate height is 1500 – 60-120= 1320 High, whilst at the sides the heights are 1300 -60 -120 =1120 high. With these overall dims you can start making the gate frame.

 A good way to ensure your gate will fit is to make a templete, in this case both halves are equal so a templete for one half is fine.

The best material to use is Hardboard. White Emulsion the smooth face and draw in normal Black Ink, the outsline and outer shape of the gate, add to this then the internal horizontal members and the vertical round infill bars (or square). From this templete you can measure your lines and create your accurate cutting lists.In the example the top vertical bars have a 130mm gap between the bars , and thus 130mm rings welded at the top, do not bother drawing the rings, The bottom vertical bars are known as 'Dog Bars' for obvious reasons.The top bars will go full height, from top to bottom, the Dog bars will be fixed at low level and at 71mm centres. It is accepted that the lower portion of gates and railings should have a gap not greater than 100mm, this to stop young 'uns getting there heads stuck! You should now be able to create a cutting list. As you get more experienced the need for a template will diminish as it is time consuming, but is a good way to learn. When complete simply roller another coat of emulsion over, let it dry and use it next time. Railings should be done using the same method.

Curved or 'Bow top' metalwork should be cut slightly longer than measured., the formed , then cut exactly to size. To create the curve shape on your template, plot the points on the vertical round bars, and use a narrow strip of flexible plastic or steel to join three or fout plotted points up, moving along until all points are joined up, the flexible material will show a natural curve rather than straight lines , point to point.

Forming Shapes

The 2 diagrams below show a 'Curving Jig', which is easily made

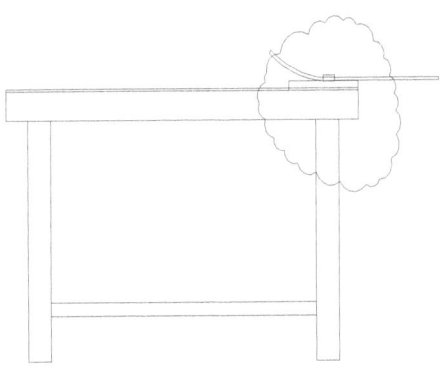

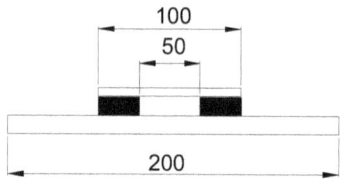

using 200 x 100 x 12mm flat for the bottom plate, 2 off 50mm x 25 x 12mm blocks (shown as shaded pieces) , with a 100 x 50 x 8mm top or 'Bridge' piece welded on top, simply mark the bars to be curved at 50mm increments and use the jig to form, surprisingly light pressure is needed to do this, the bow top gates in the Picture at the beginning of this chapter were made using this method.Commerially available bending tools can be purchased but personally I have used the method above with great success.

The method below can be used if required, this can be useful on heavier sections.

Scrolling Jigs, like the ones below can be homemade or Purchased,Personally, for commercial reasons I bought all my scrolls in, they are time consuming and not cost efficient to make, or at least I don't think so. For the hobbyist it maybe be preferable to make your own, it adds to your skillbase and can be enjoyable

90 degree angles and such like can be formed in a vice, jig or Small Handpress.

Drilling

Drilling is an important part of Steel Fabrication, it has lots of uses in Ornamental Ironwork so keeping a variety of drill bits is an absolute must, these should be kept in tip top condition, I.e not allowed to go rusty and kept sharp. If you grow as I did, I very soon bought an Ironworker which can pinch holes in thick materials very quickly, however the steel tends to distort and if you have a role of 13mm holes at 100mm centres on a 1500mm length of 30 x 8 flat it may distort, bend or twist! For me, drilling is best. Lets

look at the 3 horizontal crossmembers in the photo at the beginning of the chapter, the bottom bar has 13mm holes (to suit a 12mm bar) at 71mm centres- to allow for the dog bars., the next bar up from the bottom has the same, even though the dog bars do not go through the hole, I weld into the top

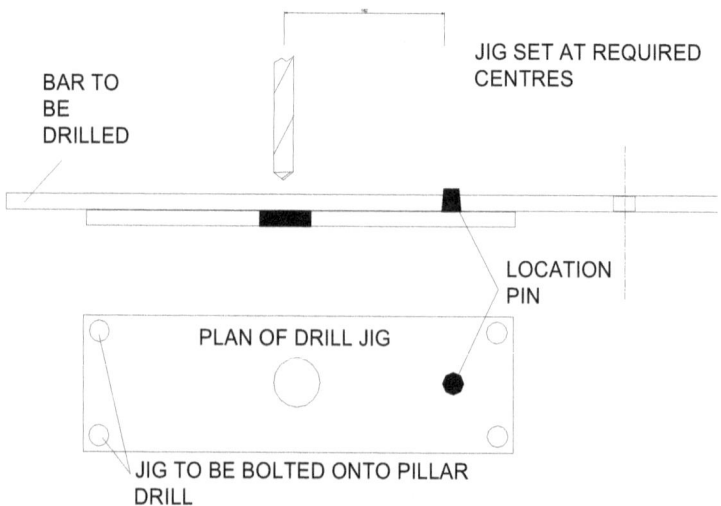

hole, it is covered with a 130mm ring in any case. The third bar has less holes at 142mm centres for the longer length bars. Some people do not drill but make the bars to fit the inside gap between the horizontal bars, but I find that drilling the bars ensure the bars not only look and are guaranteed to be parallel , the finished gate will be very much stronger.

The Drilling jog below shows an alternative method to marking out hundreds of drilling holes!

You will also need to drill various bracketry for hinges, railings, latches etc.... so safely using the correctly chosen processes will save time and give good results.

Finishing Painting, Powder Coating etc....

Dressing

I use the term 'Dressing' to mean grinding the weld, removing weld spatter, linishing and general making good all of the metal work ready for coating. In the picture below you will see the weld on the corners hasn't been linished, I grind this flat, some people prefer to leave it, either way its personal choice.'On large jobs that use heavy section, where heavy 'scale' deposits are on the surface of the steel, then subcontract 'Shot Blasting' can be desirable.

The following two pictures firstly show the one half of a set of driveway gates, the second picture shows one half 'Bi-fold' Gate , used when the span is over 14ft wide or the gates have restricted opening , usually either when you don't want to impede pedestrian access when opening outwards (pavement side) or to allow gates to be closed when car is on the drive in its parked up position(drive side).

Degreasing

Painting/Galvanising/Powder Coating

If you are intending to paint or primer the gates yourself then you should ensure that the metal surfaces are free from debris or oils and that they are clean, dry and ready to paint. A degreaser can be commercially bought where you can wipe the gates down with the solution. A subcontract Powder Coater or Painter will almost certainly have degreasing and rinsing tanks.

Some customers may ask for Galvanising prior to painting, this is known as 'hotdip galvanising' and involves dipping the gates in a tank of a molten Zinc mixture. If you intend on painting these gates you have to 'Etch Prime' the Galvanised metalwork first,a typicant suitable etch primer is generally called 'Mordant Solution'. Which you can apply by brush and wash off prior to painting.

If you decide to 'wet Spray' you need to consider proper L.E.V (Local Exhaust ventilation), an accompanying facemask of the required grade is also a must.

Some companies offer their gates as bare metal -customer hand paints themselves and save a few quid, or supplied in Primer only, this is normally Grey or Black primer, Red Oxide primer is also a common alternative.

If you decide to Wet Spray you need to ensure you have a primer and Topcoat that is compatible together, some primers wont work well with Acrylics or 'Two Pack paints' also speak to your supplier to ensure you have the correct materials, also some paints need to be 'cured' or 'baked' in an oven whereas you should be looking for 'air-drying paints'. This can be an expensive task, which I why I would always opt for PPC or Polyester Powder Coating, there are usually numerous companies ina ny town tha offer this service and you might be pleaqsantly surprised at the costs when compared to the materials , time and effort that go into wet spraying.

Highlights

Railheads and other features are sometimes highlighted in a different colour, usually Gold, Silver, Bronze or Copper

An antique Gold 'Splash' has become common, this consists of dipping a 2" brush into the paint and lightly 'swiping' the railheads for example in one swift horizontal light motion. You can see this in the picture below.

Welding

In This Section I will endeavour to cover the main type of Welding that the small Steel Fabricator will use, namely MIG welding, but will also briefly cover MMA (Stick/Arc) Welding as this sometimes will be an important method when it comes to welding in awkward areas, inaccessible to Large Mig sets or outside/Sitework situations.

I will also offer advice for the 'newbies' who are learning to weld.

Health and Safety is paramount, Personal Protective equipment should always be used as well as Fire Fighting equipment should be on hand, a bucket of sand at the very least if welding near combustible materials (i.e a Wooden Floor?, Paint , Thinners, Rags, packing materials and so on...). Along with other potential risks, please consider adequate Ventilation, Arc eye (Welding Flash), Burns, etc.... Pay particular attention to UV burns, yopu may not notice until afterwards, if you wear Short Sleeved T Shirts, or a 'V' neck Sweatshirt etc.... your skin will react in the same way as severe sun burn, i.e itchy, red and warm to touch. Rather than try and use sun tan lotion, just simply 'COVER UP'! Always use a Welding mask, even when 'Tacking Up' , use flame proof Overalls and Safety shoes or preferably boots. A leather Welding Apron and Leather 'Oversleeves' are extremely useful too as well are proper Welding Gauntlets, a welders cap will also help to stop the occasional hot spark from landing on your head! Read up on Health and Safety Equipment and practices before even venturing any further!! However, Welding can be very safe and injuries associated with Welding are perhaps more likely to be Trip hazards, Incorrect Manual lifting etc....

M.M.A Welding

Manual Metal Arc Welding also known as M.M.A, Stick and SMAW (Sheilded Metal Arc Welding) is a Manual Welding process that uses an Electrode coated in flux and an Electric Current to create a Welding Power Supply which in turn creates a welding arc between the metals to be joined. An earth cable from the Welder needs to fixed via conductible means to the workpiece, a steel bench is ideal for small items. The flux coated creates a Sheilding Gas as the welding takes place and ensures no contamination from the surround air takes place. Arc Welding leaves a trail of Molten Slag, which can normally be 'Chipped off' with a Arc Welders Hammer.

The flexibility and versatlilty that this process allows make this old technology very useful and still commonly used today. If you had to weld outdoors, MIg Welding would be difficult as the inert shielding gas used would be blown away, whereas stick welding would do the job with minimum hassle.

These days Invertor Arc sets (sometimes supplied with a Tig torch and can also be used for Tig on the more expensive end of the market) are a good buy, especially the branded ones, cheap Far Eastern ones are flooding the market and are easily available through sites like Ebay, the components are somewhat suspect and longevity for your investment is questionable, but even so a bargain can be had if drop on one for the right price. Try and avoid the cheaper Air cooled ones (sometimes called 'Buzzboxes') that discount catalogues, Automotive stores and even Superstores now offer starting at about 40-70 quid, they will cut out and blow fuses frequently, overheat and are generally harder to use, especially for the novice. Similarly, cheap Arc rods will be a poor choice, try to start with a branded one, preferably in a 'general usage' size 3.2mm. Occasionally, an old 'Oil Cooled' Arc

welder may come up for sale, these normally look like a square box slightly bigger than a large 110v Transformer but very very heavy,

great for the workshop no good for carting around with you! They are however bulletproof and last a lifetime, in fact the chances are if you find one its already 50 years + old.!

Practice on a strip of steel before trying welding steels together, get use to running an even straight line of weld down, if needed draw a straight line or chalkmark, practice is everything. Hold the torch towards you at around 15 degree from the flat piece of steel laid out horizontally and pull towards you, practice, practice practice, for the novice it will be hard at first but success will come!

Selection Table, *This works for me!*

Electrode Size	Approx Amps required	Approx. Metal Thickness
2.5	50-100	1.5-2.00
3.2	100-130	2.00-5.00
4.0	130-160	5.00-8.00

5.0	160-275	8.00-10.00

Thicker Rods or Multi runs are possible for thicker materials.

I find An Arc welder invaluable when site fitting gates and railings, usually for welding new Hinges to existing Steel posts, or Welding realigned latches on, Refurb jobs etc....Thus, A very strong and permanent weld can be effected with minimal skill in a cost efficient way.

Note: For workshop use, and on Steelwork in general, Stick Welding is *slower and more expensive* than Mig!

Mig Welding (GMAW -Gas Metal Arc Welding)

In any steel Fabrication shop , whether a specialist Structural SteelFabricator, Sheetmetal Workshop, Heavy Plate work,Production Welding shop or bespoke Fabrications you will find a Mig Welder,' M.i.g' means *'Metal Inert Gas'.* In this case (for steel) the Electrode is usually a copper coated steel wire, usually bought on a 15kg reel, a shielding gas such as a CO_2 /Argon mix is most often used , each Gas supplier will have their own brand name ands sometimes recommend a different gas depending on the Thickness of Steel to be welded. A gauge is fixed to the Welding bottle that shows the gas flow rate and bottle pressue. There are also specialist wires for other materials such as Stainless Steel and Aluminium, also for some Mig Sets Flux cored wires, but generally not used in general Steel Fabrication, more for the hobbyist on 'No Gas' type welders.

The wire will Usually be 1.00 thick for general use, but 0.6, 0.8 and 1.2mm thick are other common sizes. With Mig welding you will generate lots of weld spatter which can sometimes be hard to remove, use a good quality Anti – Spatter spray to spray the workpiece, 'Tip-Dip' or similar can be used to immerse the welding tip in, this will help prevent Spatter from sticking to the inside of the Shroud and possibly restricting gas flow shortening the life of the consumables such as Tips, Shroud etc....

How do I Weld?

Do I just Just squeeze the trigger and away i go ? Not quite, but its quite simple to learn but not so simple to master, however, quality and quantity will improve quickly with regular practice.

Practice welding straight beads first before you attempt joining metal, before pulling the trigger leave about 6-10 mm of wire sticking out of the tip, tilt the torch at about 10 degrees. For ease of use, use a reactive type headshield so both hands are free, use one hand to steady as you weld , especially whilst a novice, also ensure that your lense is the correct shade and you can see easily where you intend to weld.

Whilst Mig Welders will all have their own personal preference, regarding the Push or Pull method, you can actually weld either with great success, pulling the torch towards you will give more penetration but a narrower weld bead, pushing will give reduced (but usually ample) penetration and a wider weld bead. Positioning your workpiece on an incline and welding downhill will both speed up and improve your welding and is a good way to practice, basically the molten flow is easier to control.

Remember to keep your torch cable as straight as you can, any bends can cause the wire to drag and not exit the torch at a consistent speed meaning intermittent and poor welding.

Welding Parameters.

Setting your welder for different types and thicknesses of steels is the first thing you need to master, once you know the basics it's just a matter of getting to know your welder.

Wire Speeds, generally control Amperage, Thicker Material= increased Voltage = Increase Wire Speed. Like the Voltage dials these are normally numbered and after regular use the operator will quickly get used to these settings. The voltage dials normally consist of 2 dials , one for a 'coarse' voltage, numbered 1-5 for example and a 2nd dial used as a 'Fine' voltage controller similarly numbered.

Depending on the actual operator, and how quickly or slowly they move the torch, the settings used will vary from Operator to Operator. It has been said to me that Mig welding that sounds like 'Sizzling Bacon' indicates your on the right track when learning. A pretty good analogy I reckon.

Troubleshooting

Lack of Fusion – ensure Welding parameters are set properely, sometimes hard to visually detect a problem so a simple destructive test may show the problem, sometimes the consumable as failed to adhere to the base or parent metal.

Wirefeed problem-This may be because of a faulty liner inside the torch cable, dirty or fouled tip stopping the wire from leaving the tip cleanly, incorrect adjustment of the spool holder – too tight or too loose, Wire rollers incorrect ones used – the rollers are usually stamped 0.8 for example, using 0.8 wire with 1.2mm wire will cause the wire to possibly 'slip'.

Porosity Problem- Firstly check that the metal is free from paint, rust or oil etc...This is a common and easy to overlook problem. The next things to check are that the sheilding gas is flowing correctly, sometimes its easy to forget to turn the gas bottle valve on!, also make sure the gas bottle to welder tubes are free and not coiled. Are you welding in a draft or wind laden atmosphere? This will cause the sheilding gas to be blown away and porous welds. Lastly, if you are using the pull technique try the push method instead, this will ensure the weld pool is covered by the Sheilding gas.

Incorrect Torch Position- I choose to lean the torch 10-15 degrees and keep the wire about 6-10 mm from the metal, i've found (but everyones different!) the push method will normally give best results. This is a regular problem for the novice, this and welding too slow or too fast, only practice will determine what is best method for you.

WELD SYMBOLS

SQUARE BUTT WELD	SINGLE V BUTT WELD	SINGLE BEVEL BUTT WELD
SINGLE-U BUTT WELD	SINGLE-J BUTT WELD	BACKING RUN
FILLET WELD	PLUG WELD	SPOT WELD

CHAPTER 9...... *Some Examples:*

just

another

example

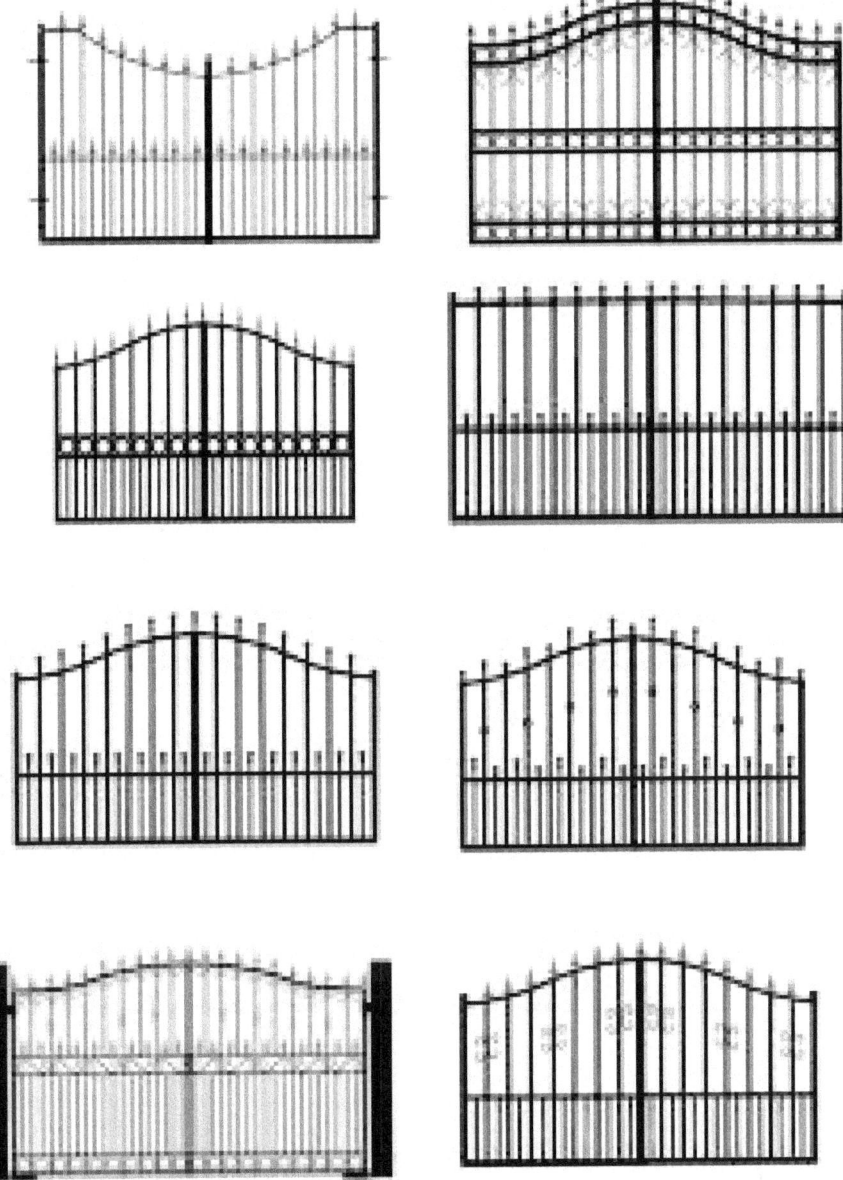

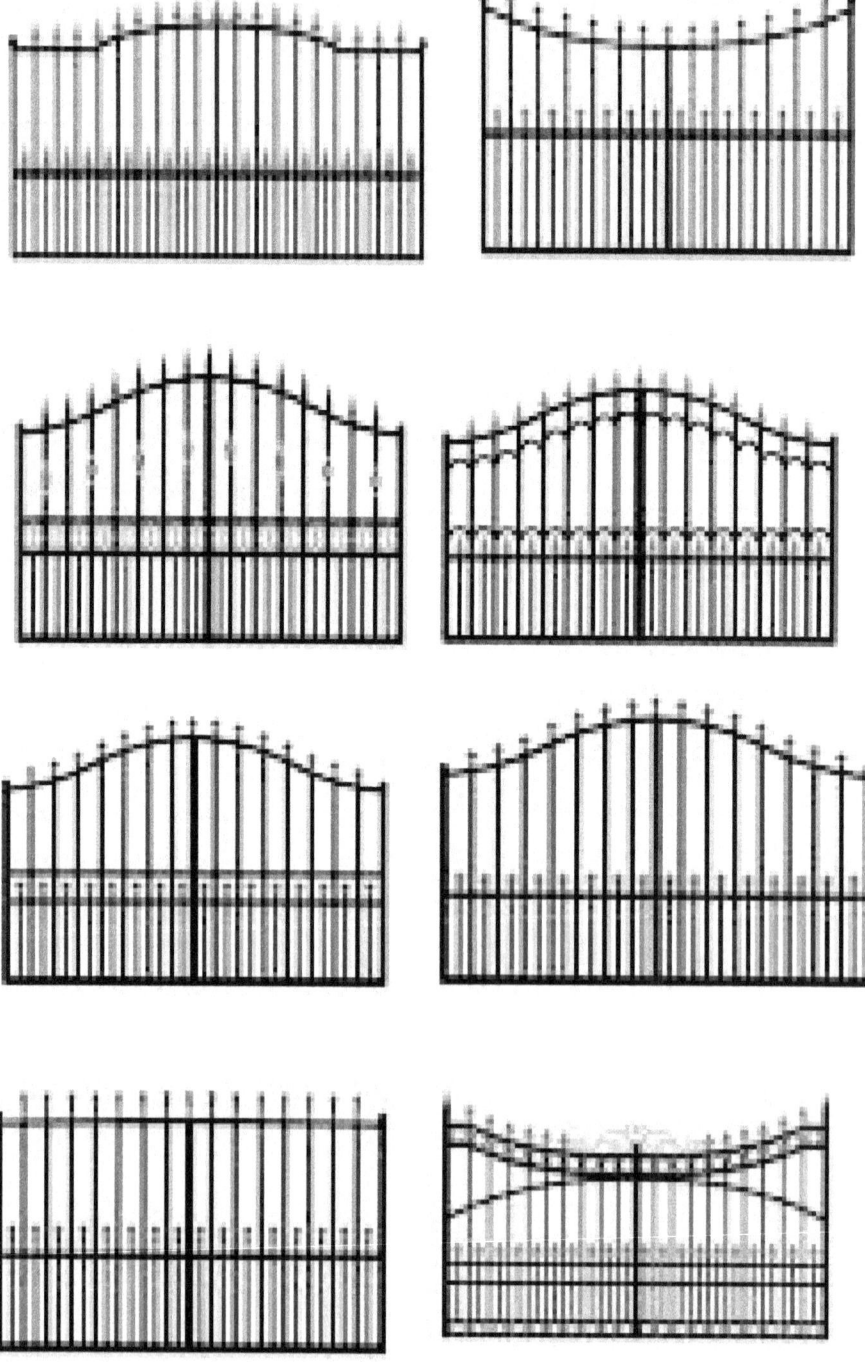

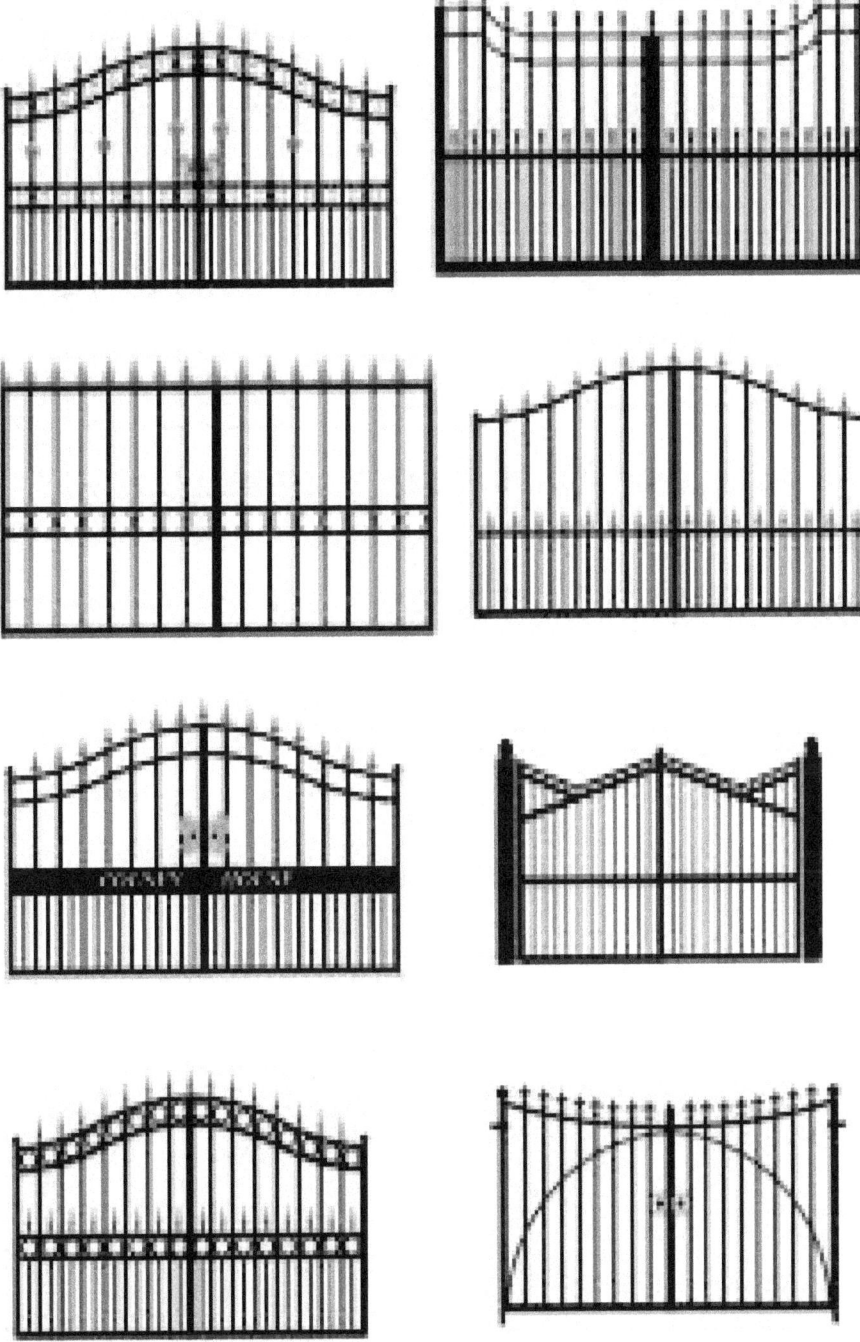

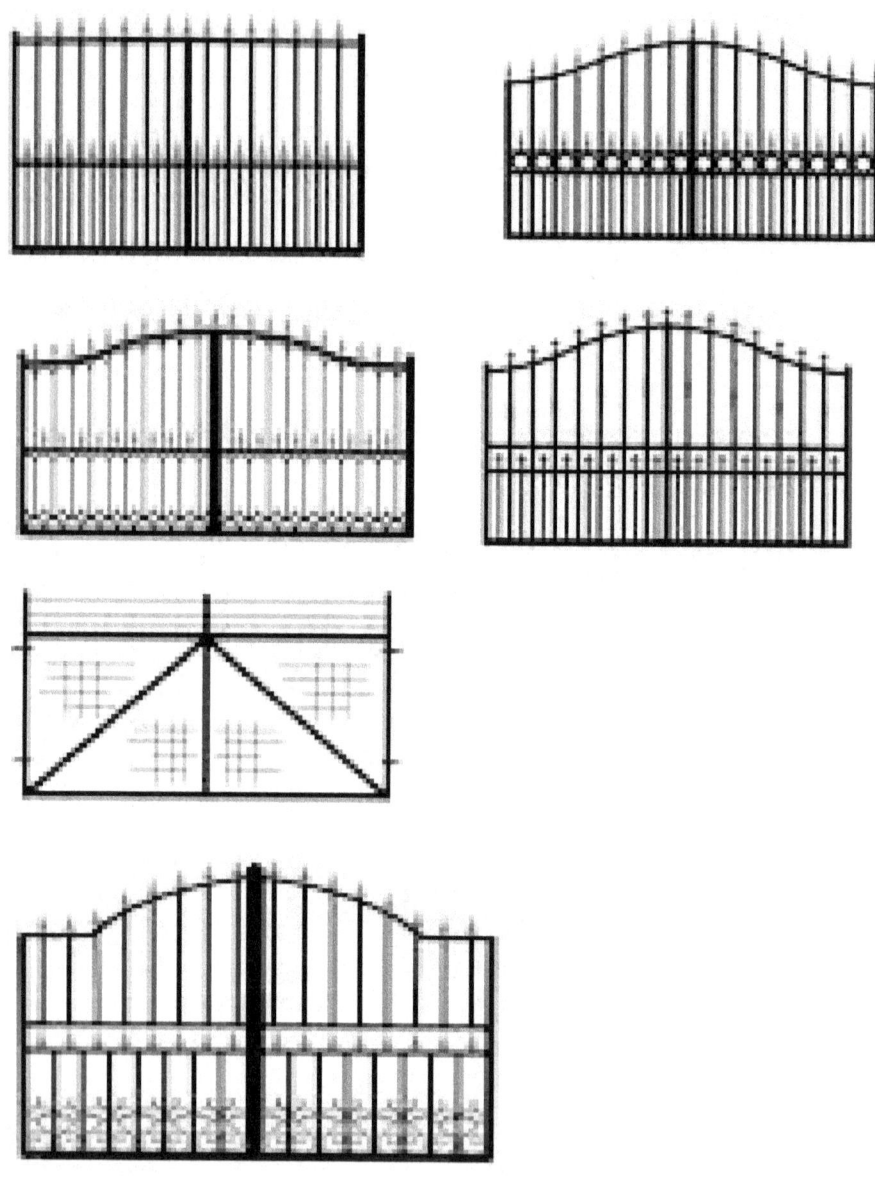

Photo 1: Railings, Complete with Fixed Posts

Photo 2: Bi-Fold double gates

Photo 3: Railings to match the above

Photo 4: BUDGET RAILINGS, 1METRE HIGH WITH SILVER BALL TOPS

Photo 5: STAGGERED 1500MM HIGH RAILINGS

Photo6: DRIVE GATES - MY FIRST ATTEMPT IN THE SHED, CIRCA 1990

Photo 7: 16MM SQUARE TWISTED BAR RAILINGS -HEAVY DUTY

Photo 8: SMALL
SINGLE FLEUR-DE-LYS -
KENSINGTON LONDON

Photo 9: RAILINGS WITH
HALF ROUND INFILLS

Photo 10: BUDGET BI-FOLD DOUBLE GATES C/W SINGLE
GATE AND RAILS

Photo 12: BUDGET SINGLE

Photo 13: BUDGET TALL SINGLE

Photo 14: SHOWS 'ANTIQUE ' GOLD 'SPLASH' ON BLACK RAILHEADS

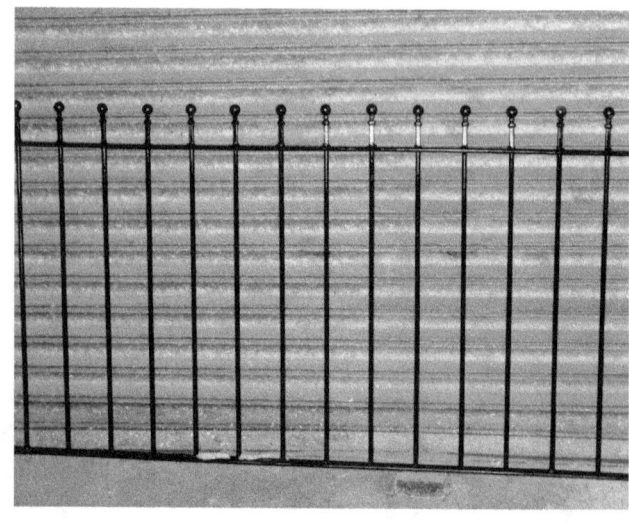

e.g's of Other Fabrication Work

Steel Doors

Mild steel Welding

Stainless welding